INTERNATIONAL CENTRE FOR MECHANICAL SCIENCES

COURSES AND LECTURES - No. 73

BERNARD D. COLEMAN
CARNEGIE - MELLON UNIVERSITY, PITTSBURGH

# THERMODYNAMICS OF MATERIALS WITH MEMORY

COURSE HELD AT THE DEPARTMENT
OF MECHANICS OF SOLIDS
JULY 1971

UDINE 1971

SPRINGER-VERLAG WIEN GMBH

Originally published by Springer-Verlag Wien-New York in 1972

ISBN 978-3-211-81125-2 ISBN 978-3-7091-2951-7 (eBook)

DOI 10.1007/978-3-7091-2951-7

This text has been submitted as being "ready for camera" by professor B.D. Coleman and has been reproduced without any corrections or additions, except for the titles at the top of pages.

For the CISM.

Dr. Franco Buttazzoni

Responsable for the Editorial Board

## 1. Introduction

In theories of the dynamical behavior of continua, there are several ways of describing the dissipative effects which, in addition to heat conduction, accompany deformation. The oldest way is to employ a viscous stress which depends on the rate of strain, as is done in the theory of Navier-Stokes fluids. In another description of dissipation, one postulates the existence of internal state variables which influence the stress and obey differential equations in which the strain appears. A third approach is to assume that the entire past history of the strain influences the stress in a manner compatible with a general postulate of smoothness or "principle of fading memory".

Experience in high-polymer physics shows that the mechanical behavior of many materials, including polymer melts and solutions, as well as amorphous, cross-linked solids and semi-crystalline plastics, is more easily described within the theory of materials with fading memory than by theories of the viscous-stress type, which do not account for gradual stress-relaxation, or by theories which rest on a finite number of internal state variables and which, therefore, give rise to discrete relaxation spectra when linearized.

Some years ago, Walter Noll and I proposed a systematic procedure for rendering explicit the restrictions which the second law places on constitutive relations.[1] The procedure was easily applied in theories of materials of the viscous-stress type[1,2] and in theories which employ evolution equations for internal state variables[3]. These applications, which did not yield results a physicist would consider surprising, were presented as attempts at clarification, with the emphasis laid upon logical relations. Implementation of the procedure in the theory of materials with memory was a different matter, however, for it there led to conclusions[4] which, although not anticipated by other arguments, have recently been shown to have important bearing on wave propagation[5] and dynamical stability[6,7] In this course we shall discuss the restrictions which the second law places on the response functionals of materials with memory. Although it is possible to develop analogous theories for materials with "permanent memory",[#] here we shall emphasize materials which possess "fading memory" in the sense that configurations experienced in the recent past have a stronger influence on the present values of the stress and free energy than configurations experienced in the distant past.

[#]Cf. Owen's discussion of the thermodynamics of materials with elastic range,[8] Owen and Williams' theory of rate-independent materials,[9] and a recent essay,[10] in which Owen and I generalize the present treatment.

## 2. Constitutive Assumptions and the Second Law

Let a fixed reference configuration $\mathcal{R}$ be assigned for the body $\mathcal{B}$ under consideration, and identify each of the material points X of $\mathcal{B}$ with the place $\underset{\sim}{\xi}$ in space that X occupies when $\mathcal{B}$ has the configuration $\mathcal{R}$. A <u>thermodynamic</u> <u>process</u> of $\mathcal{B}$ is a collection of functions of $\underset{\sim}{\xi}$ and time compatible with the laws of balance of momentum and energy. For the materials covered by the present theory, each process consists of eight functions: (1) the <u>motion</u> $\underset{\sim}{\chi}$, with $\underset{\sim}{x} = \underset{\sim}{\chi}(\underset{\sim}{\xi},t)$ called the <u>position</u> at time t of the material point located at $\underset{\sim}{\xi}$ in $\mathcal{R}$, (2) the local absolute <u>temperature</u> $\theta$, which is assumed to be positive, (3) the symmetric <u>stress</u> <u>tensor</u> $\underset{\sim}{T}$ of Cauchy, (4) the specific <u>internal</u> <u>energy</u> $\epsilon$, per unit mass, (5) the specific <u>entropy</u> $\eta$, per unit mass, (6) the <u>heat</u> <u>flux</u> <u>vector</u> $\underset{\sim}{q}$, (7) the <u>body</u> <u>force</u> $\underset{\sim}{b}$, per unit mass (exerted on $\mathcal{B}$ at $\underset{\sim}{x} = \underset{\sim}{\chi}(\underset{\sim}{\xi},t)$ by the "external world", i.e. by other bodies which do not intersect $\mathcal{B}$), and (8) the <u>rate</u> <u>of</u> <u>heat</u> <u>supply</u> r (i.e. the radiation energy, per unit mass and unit time, absorbed by $\mathcal{B}$ at $\underset{\sim}{x} = \underset{\sim}{\chi}(\underset{\sim}{\xi},t)$ and furnished by the "external world"). The first six of these functions determine the process, for once $\underset{\sim}{\chi}$, $\theta$, $\underset{\sim}{T}$, $\epsilon$, $\eta$, and $\underset{\sim}{q}$ have been specified for all $\underset{\sim}{\xi}$ and t, the functions $\underset{\sim}{b}$ and r are determined[1] by the requirement that the process obey the laws of balance of momentum and energy, which state that, for each part $\mathcal{P}$ of $\mathcal{B}$ and each time t,

$$\frac{d}{dt}\int_{\mathcal{P}} \underset{\sim}{\dot{x}}\, dm = \int_{\mathcal{P}} \underset{\sim}{b}\, dm + \int_{\partial\mathcal{P}} \underset{\sim}{T}\underset{\sim}{n}\, da \qquad (2.1)$$

and

$$\frac{d}{dt}\int_{\mathcal{P}}\left(\epsilon+\frac{1}{2}\dot{\underset{\sim}{x}}\cdot\dot{\underset{\sim}{x}}\right)dm = \int_{\partial\mathcal{P}}(\dot{\underset{\sim}{x}}\cdot\underset{\sim}{b}+r)dm + \int_{\partial\mathcal{P}}(\dot{\underset{\sim}{x}}\cdot\underset{\sim}{T}\underset{\sim}{n}-\underset{\sim}{q}\cdot\underset{\sim}{n})da. \qquad (2.2)$$

In these equations dm is the element of mass in the body, $\partial\mathcal{P}$ is the surface of $\mathcal{P}$ in the configuration at time t, da is the element of surface area, $\underset{\sim}{n}$ is the exterior unit normal vector to $\partial\mathcal{P}$ and the superposed dots denote material time-derivatives.

The specific free energy $\psi$ (also called the "Helmholtz free energy per unit mass") is defined by

$$\psi = \epsilon - \theta\eta. \qquad (2.3)$$

The deformation gradient $\underset{\sim}{F}$ is the gradient of $\underset{\sim}{\chi}(\underset{\sim}{\xi},t)$ with respect to $\underset{\sim}{\xi}$:

$$\underset{\sim}{F} = \underset{\sim}{F}(\underset{\sim}{\xi},t) = \nabla_{\underset{\sim}{\xi}}\underset{\sim}{\chi}(\underset{\sim}{\xi},t). \qquad (2.4)$$

It is assumed that $\underset{\sim}{F}$ is non-singular; hence

$$\det \underset{\sim}{F} \neq 0. \qquad (2.5)$$

The Piola-Kirchoff tensor, $\underset{\sim}{S} = \underset{\sim}{S}(\underset{\sim}{\xi},t)$, is defined by

$$\underset{\sim}{S} = \frac{1}{\rho}\underset{\sim}{T}\underset{\sim}{F}^{T^{-1}}, \quad \text{i.e. } \rho\underset{\sim}{S}\underset{\sim}{F}^{T} = \underset{\sim}{T}, \qquad (2.6)$$

with $\rho$ the mass density. I denote by $\underset{\sim}{g}$ the spatial gradient of the temperature, i.e. the gradient of $\theta$ considered a function of the present position $\underset{\sim}{x} = \underset{\sim}{\chi}(\underset{\sim}{\xi},t)$:

$$\underset{\sim}{g} = \nabla_{\underset{\sim}{x}}\theta\left(\underset{\sim}{\chi}^{-1}(\underset{\sim}{x},t),t\right), \quad \text{or} \quad \underset{\sim}{F}^{T}\underset{\sim}{g} = \nabla_{\underset{\sim}{\xi}}\theta(\underset{\sim}{\xi},t). \qquad (2.7)$$

Let $\underset{\sim}{F}(\tau)$ and $\theta(\tau)$ be the deformation gradient and temperature at time $\tau$ at a fixed material point X. The functions $\underset{\sim}{F}^t$ and $\theta^t$, defined by

$$\underset{\sim}{F}^t(s) = \underset{\sim}{F}(t-s), \qquad \theta^t(s) = \theta(t-s), \qquad 0 \le s < \infty, \tag{2.8}$$

are called the <u>history up to</u> t <u>of the deformation gradient at</u> X and the <u>history up to</u> t <u>of the temperature at</u> X; $\underset{\sim}{F}^t$ maps $[0,\infty)$ into the set of non-singular tensors, while $\theta^t$ maps $[0,\infty)$ into the set of positive numbers.

Each material is characterized by constitutive relations which limit the class of processes possible in a body comprised of the material. In the thermodynamics of materials with memory, a <u>simple material</u> is one for which the free energy, the stress, the entropy, and the heat flux are determined when the history of the deformation gradient, the history of the temperature, and the present value of the temperature gradient are specified.[4] Thus, at each material point of a simple material there hold equations of the form

$$\left.\begin{aligned} \psi(t) &= \mathfrak{p}(\underset{\sim}{F}^t,\theta^t;\underset{\sim}{g}(t)), \\ \eta(t) &= \mathfrak{h}(\underset{\sim}{F}^t,\theta^t;\underset{\sim}{g}(t)), \\ \underset{\sim}{S}(t) &= \mathfrak{s}(\underset{\sim}{F}^t,\theta^t;\underset{\sim}{g}(t)), \\ \underset{\sim}{q}(t) &= \mathfrak{q}(\underset{\sim}{F}^t,\theta^t;\underset{\sim}{g}(t)). \end{aligned}\right\} \tag{2.9}$$

It is assumed that the four functions $\mathfrak{p}$, $\mathfrak{h}$, $\mathfrak{s}$, and $\mathfrak{q}$ are given at each material point; these functions, called "response functionals" or "constitutive functionals" depend, of course, on the choice of the reference configuration.# A process is said to be admissible in the simple material if, in addition to obeying the balance laws (2.1) and (2.2), it obeys the constitutive relations (2.9).

If one regards $\underset{\sim}{q}/\theta$ to be a vectorial flux of entropy and $r/\theta$ to be a scalar supply of entropy, then it is natural to define the rate of production of entropy in a part $\mathcal{P}$ of $\mathcal{B}$ to be

$$\Gamma(\mathcal{P},t) = \frac{d}{dt}\int_{\mathcal{P}} \eta \, dm - \left[\int_{\mathcal{P}} \frac{r}{\theta}\, dm - \int_{\partial\mathcal{P}} \underset{\sim}{q}\cdot\underset{\sim}{n}\, da\right]. \tag{2.10}$$

The Clausius-Duhem inequality[12] is the assertion that

$$\Gamma(\mathcal{P},t) \geq 0. \tag{2.11}$$

In our paper[1] of 1963, Noll and I pointed out that in many branches of continuum physics the second law of thermodynamics can be given a precise mathematical meaning if it is interpreted to be the following principle.

Dissipation Principle. For every admissible thermodynamic process in a body $\mathcal{B}$, the Clausius-Duhem inequality (2.11) must hold at all times $t$ and in all parts $\mathcal{P}$ of $\mathcal{B}$.

It is clear that this principle implies that response functionals cannot be chosen arbitrarily. In Section 3 I shall list the restrictions

#Cf. Noll[11].

which the principle places on $\mathfrak{p}$, $\mathfrak{h}$, $\mathfrak{s}$, and $\mathfrak{q}$ in (2.9) when these functionals obey the postulate of regularity called the "principle of fading memory".# First, however, I should like to outline a recently developed axiomatic approach[13] to the theory of fading memory.

#For earlier studies of the principle of fading memory, see Refs. 14-18. Truesdell & Noll[19] have surveyed the work done up to 1965.

## 3. Fading Memory

Let us use Greek majuscules, such as $\Lambda$, to denote ordered pairs $(\underset{\sim}{L},\lambda)$, with $\underset{\sim}{L}$ a tensor and $\lambda$ a scalar. The definitions

$$\left.\begin{aligned} \alpha\Lambda_1 + \beta\Lambda_2 &= \alpha(\underset{\sim}{L}_1,\lambda_1) + \beta(\underset{\sim}{L}_2,\lambda_2) = (\alpha\underset{\sim}{L}_1 + \beta\underset{\sim}{L}_2,\ \alpha\lambda_1 + \beta\lambda_2), \\ \Lambda_1\cdot\Lambda_2 &= \mathrm{tr}(\underset{\sim}{L}_1\underset{\sim}{L}_2^T) + \lambda_1\lambda_2 \end{aligned}\right\} \tag{3.1}$$

make the set of all such ordered pairs a 10-dimensional vector space $\mathcal{V}_{(10)}$ with norm

$$|\Lambda| = \sqrt{\Lambda\cdot\Lambda} = \sqrt{\mathrm{tr}(\underset{\sim}{L}\underset{\sim}{L}^T) + \lambda^2}. \tag{3.2}$$

The elements of $\mathcal{V}_{(10)}$ of the type

$$\Gamma = (\underset{\sim}{F},\theta), \tag{3.3}$$

with $\underset{\sim}{F}$ a non-singular tensor and $\theta$ a positive number, form a cone $\mathcal{C}$ in $\mathcal{V}_{(10)}$. (A subset $\mathfrak{U}$ of a vector space is called a <u>cone</u> if $\underset{\sim}{u}\in\mathfrak{U}$ and $b>0$ imply $b\underset{\sim}{u}\in\mathfrak{U}$.)

At a given material point in a process, the <u>total history up to</u> t, i.e. the history up to t of the deformation gradient and temperature, is the function $\Gamma^t = (\underset{\sim}{F}^t,\theta^t)$, mapping $[0,\infty)$ into $\mathcal{C}$:

$$\Gamma^t(s) = \left(\underset{\sim}{F}^t(s),\theta^t(s)\right) \qquad \text{for} \qquad 0 \le s < \infty. \tag{3.4}$$

The ordered pair $\Sigma$, defined by

$$\Sigma = (\underset{\sim}{S},-\eta) = \left(\frac{1}{\rho}\underset{\sim}{T}\underset{\sim}{F}^{T^{-1}},-\eta\right)\in\mathcal{V}_{(10)}, \tag{3.5}$$

is called the <u>stress-entropy vector</u>.[4] If one writes simply $\psi$ for $\psi(t)$, $\boldsymbol{g}$ for $\boldsymbol{g}(t)$, and $\Sigma$ for $\Sigma(t)$, the constitutive equations (2.9) become, in the present notation,

$$\left.\begin{aligned} \psi &= \mathfrak{p}(\Gamma^t;\boldsymbol{g}), \\ \Sigma &= \mathfrak{S}(\Gamma^t;\boldsymbol{g}), \\ \boldsymbol{q} &= \mathfrak{q}(\Gamma^t;\boldsymbol{g}), \end{aligned}\right\} \qquad (3.6)$$

where the response functional $\mathfrak{S}$ has the "components"

$$\mathfrak{S} = (\mathfrak{s}, -\mathfrak{h}). \qquad (3.7)$$

It is frequently possible to prove theorems in a branch of continuum physics without specifying the form of response functionals, but usually one must assume something about their smoothness. For this reason several topologies have been proposed as appropriate for sets of histories.[14-18,4,8,9,10]

Let us suppose that the histories $\Lambda^t$ of interest form a cone $\mathfrak{C}$ in a Banach function-space $\mathfrak{B}$. Certain basic, but usually tacit, assumptions of physical theories place limitations on the choice of the function space $\mathfrak{B}$ and its norm $\|\cdot\|$. I list below three of these requirements.#

(1) Given an arbitrary history $\Gamma^t$ in the domain D of a constitutive functional## and a positive number $\sigma$, one expects to find in D the history $\Gamma^{t+\sigma}$ for processes in which $\Gamma = (\boldsymbol{F}, \theta)$ (at some fixed

#Cf. Coleman & Mizel[18,13].

##D is here the domain for a fixed value of $\boldsymbol{g}$.

material point) has the history $\Gamma^t$ up to time t and is constant throughout the interval $[t, t+\sigma]$. The history $\Gamma^{t+\sigma}$ in such a process is called the "static continuation of $\Gamma^t$ by the amount $\sigma$". The static continuation of a history should be well defined even if one identifies the history with the set of functions at zero distance from it in $\mathfrak{B}$.

(2) If the history $\Gamma^t$ of $\Gamma$ up to time t is in the domain D, then one expects to find in D the histories $\Gamma^{t-\sigma}$ of $\Gamma$ up to previous times $t-\sigma$, $\sigma \geq 0$. These earlier histories are called "$\sigma$-sections of $\Gamma^t$".

(3) Since it should be possible to evaluate response functionals at "equilibrium states", one expects D to contain constant histories of the form $\Gamma^t(s) \equiv \Omega$, $0 \leq s < \infty$.

Victor Mizel and I have found some apparently useful implications of these elementary physical requirements, and I summarize below some of our results.[13]

Let $\mu$ be an <u>influence measure</u>; that is, a non-trivial (i.e. not identically zero), sigma-finite, positive, regular Borel measure on $[0,\infty)$ and let $\mathcal{A}$ be the set of all $\mu$-measurable functions $\phi$ mapping $[0,\infty)$ into $[0,\infty)$. Let $\nu$ be a function on $\mathcal{A}$ such that for all $\phi$ (or $\phi_i$) in $\mathcal{A}$:

(i) $0 \leq \nu(\phi) \leq \infty$, and $\nu(\phi) = 0$ if and only if $\phi(s) = 0$ $\mu$-a.e.[#];

(ii) $\nu(\phi_1 + \phi_2) \leq \nu(\phi_1) + \nu(\phi_2)$, and $\nu(a\phi) = a\nu(\phi)$ for all numbers $a \geq 0$;

[#] i.e. for all s in $[0,\infty)$ except for a set Z with $\mu(Z) = 0$.

(iii) if $\phi_1(s) \leq \phi_2(s)$ $\mu$-a.e., then $\nu(\phi_1) \leq \nu(\phi_2)$;

(iv) there is at least one function $\psi$ in $\mathcal{A}$ with $0 < \nu(\psi) < \infty$;

(v) if $\psi, \phi_1, \phi_2, \ldots$ are in $\mathcal{A}$ and if $\phi_n(s) \uparrow \psi(s)$ $\mu$-a.e., then $\nu(\phi_n) \uparrow \nu(\psi)$.

Such a function $\nu$ is called a non-trivial function norm, relative to $\mu$, with the sequential Fatou property.#

Let $\bar{V}$ be the set of $\mu$-measurable functions mapping $[0,\infty)$ into $\mathcal{V}_{(10)}$, and let $\|\cdot\|$ be the function on $\bar{V}$ defined by

$$\|\Phi\| = \nu(|\Phi|) \tag{3.8}$$

for each $\Phi$ in $\bar{V}$. I write V for the set of all $\Phi$ in $\bar{V}$ with $\|\Phi\| < \infty$. Each function $\Phi$ in V is called a history, and its independent variable (usually denoted by s) is called the elapsed time. The value $\Phi(0)$ of $\Phi$ at $s = 0$ is called the present value of $\Phi$, and the past values are those for which $0 < s < \infty$. The function space $\mathfrak{B}$ obtained by calling two functions $\Phi_1$ and $\Phi_2$ in V the same whenever $\|\Phi_1 - \Phi_2\| = 0$ is easily shown to be a Banach space; it is called a history space or, at length, a Banach function space formed from histories with values in $\mathcal{V}_{(10)}$.

Let C be the class of functions $\Phi$ in V such that $\Phi(s)$ is in $\mathcal{C}$ for all $s \geq 0$, and let $\mathfrak{C}$ be the set of equivalence classes obtained by calling the same those elements $\Phi_1$, $\Phi_2$ of C for which $\|\Phi_1 - \Phi_2\| = 0$.

---

#Cf. Luxemburg & Zaanen[21] and the literature quoted by them.

Clearly, C is a cone in V, and $\mathfrak{C}$ is a cone contained in the Banach function-space $\mathfrak{B}$. Let $\mathfrak{C}$ be the domain D of definition of the response functionals in (3.6). (When the dependence on $\underset{\sim}{g}$ is under discussion, the domain is taken to be the set $\mathfrak{C} \times \mathcal{V}_{(3)}$, which forms a cone in $\mathfrak{B} \oplus \mathcal{V}_{(3)}$.)

If $\Psi$ is a function on $[0,\infty)$ and $\sigma$ a positive number, then the <u>static continuation</u> of $\Psi$ by the amount $\sigma$ is the function $\Psi^{(\sigma)}$ on $[0,\infty)$ defined by[16]

$$\Psi^{(\sigma)}(s) = \begin{cases} \Psi(0), & 0 \leq s < \sigma, \\ \Psi(s-\sigma), & \sigma < s < \infty, \end{cases} \tag{3.9}$$

and the $\sigma$-<u>section</u> of $\Psi$ is the function $\Psi_{(\sigma)}$ on $[0,\infty)$ given by[18]

$$\Psi_{(\sigma)}(s) = \Psi(s+\sigma), \qquad 0 \leq s < \infty. \tag{3.10}$$

If $\Psi$ is the history up to t of $\Gamma = (\underset{\sim}{F},\theta)$ (at a fixed material point X in some particular process), then $\Psi_{(\sigma)}$ is the history of $\Gamma$ up to $t-\sigma$, while $\Psi^{(\sigma)}$ gives the history of $\Gamma$ up to $t+\sigma$ assuming that $\Gamma$ is held constant from t to $t+\sigma$. The physical requirements (1) and (2) stated above are made precise by laying down the following two postulates.[18,13]

**Postulate 1.** <u>If a given function $\Phi$ is in C, then all its static continuations $\Phi^{(\sigma)}$, $\sigma \geq 0$, are also in C. Furthermore, if $\Phi$ and $\Psi$ in C are such that $\|\Phi-\Psi\| = 0$, then $\|\Phi^{(\sigma)} - \Psi^{(\sigma)}\| = 0$ for all $\sigma \geq 0$.</u>

**Postulate 2.** <u>If $\Phi$ is in C, then so also are all its $\sigma$-sections, $\Phi_{(\sigma)}$, $\sigma \geq 0$.</u>

Employing Postulate 1, one can easily prove the following theorem which shows that the present value $\Phi(0)$ of a history $\Phi$ has a special status, in the sense that the norm $\|\Phi\| = \nu(|\phi|)$ places greater emphasis on $\Phi(0)$ than on any individual past value.

**Theorem 1.**[#] *The influence measure $\mu$ must have an atom at $s = 0$ and be absolutely continuous on $(0,\infty)$ with respect to Lebesgue measure.*

Postulates 1 and 2, together, yield

**Theorem 2.**[##] *Either $\mu((0,\infty)) = 0$, or Lebesgue measure is absolutely continuous on $(0,\infty)$ with respect to $\mu$.*

Thus the $\mu$-measure of the singleton $\{0\}$ is not zero, and if $\mu((0,\infty))$ is not zero, then an arbitrary subset of $(0,\infty)$ has zero $\mu$-measure if and only if it has zero Lebesgue measure. So as to have a non-trivial theory, *let us assume that $\mu((0,\infty))$ is not zero*. Since the measure $\mu$ is employed here only to render precise the expression "$\mu$-a.e." in the axioms (i), (iii), (iv), and (v) for $\nu$, Theorems 1 and 2 imply that one can here replace $\mu$ with the Borel measure on $[0,\infty)$ that assigns the value 1 to the singleton $\{0\}$ and equals Lebesgue measure when restricted to Borel subsets of $(0,\infty)$.

---

[#] Ref. 13, Thm. 2.1.

[##] Ref. 13, Thm. 2.2.

If $\Phi$ is a function in V, the restriction of $\Phi$ to $(0,\infty)$ is called the past history of $\Phi$ and is denoted by $\Phi_r$. Let $V_r$ be the set of all functions $\Phi_r$ obtained by restricting members of V to $(0,\infty)$, and define $\|\cdot\|_r$ on $V_r$ by

$$\|\Phi_r\|_r = \|\Phi\chi_{(0,\infty)}\|, \tag{3.11}$$

with $\chi_{(0,\infty)}$ the characteristic function[#] of $(0,\infty)$. The space of past histories is the function space $\mathfrak{B}_r$ obtained by calling the same those elements $\Phi_r$, $\Psi_r$ of $V_r$ for which $\|\Phi_r - \Psi_r\|_r = 0$. It is easily verified that $\|\cdot\|_r$ is a norm on $\mathfrak{B}_r$ and that $\mathfrak{B}_r$ is a Banach space. I write $C_r$ for the set of functions in $V_r$ with values in $\mathcal{C}$ and $\mathfrak{C}_r$ for the corresponding cone in $\mathfrak{B}_r$.

An immediate consequence of Theorems 1 and 2 is

Theorem 3.[##] $\mathfrak{B} = \mathcal{V}_{(10)} \oplus \mathfrak{B}_r$, *and the norm* $\|\cdot\|$ *on* $\mathfrak{B}$ *is equivalent to the norm* $\|\cdot\|'$ *defined by*

$$\|\Phi\|' = |\Phi(0)| + \|\Phi_r\|_r. \tag{3.12}$$

Here $|\cdot|$ is the original norm (3.2) on $\mathcal{V}_{(10)}$, $\|\cdot\|$ is the norm on $\mathfrak{B}$ defined in (3.8), and $\|\cdot\|_r$ is the norm on $\mathfrak{B}_r$ defined in (3.11). The equivalence of $\|\cdot\|'$ and $\|\cdot\|$ means that there exist two positive numbers

---

[#] i.e., $\chi_{(0,\infty)}$ has domain $[0,\infty)$ and is such that $\chi_{(0,\infty)}(s) = 1$ for $s \in (0,\infty)$, while $\chi_{(0,\infty)}(0) = 0$.

[##] Ref. 13, Thm. 3.1.

$c_1$ and $c_2$ such that

$$c_1\|\Phi\| \leq \|\Phi\|' \leq c_2\|\Phi\|$$

for all $\Phi$ in $\mathfrak{B}$. It follows from Theorem 3 that, even after the functions in V are grouped together to form the equivalence classes comprising $\mathfrak{B}$, each history $\Phi$ has a well-defined present value $\Phi(0)$.

If $\Omega$ is a vector, $\Omega^\dagger$ denotes the constant function on $[0,\infty)$ with value $\Omega$:

$$\Omega^\dagger(s) = \Omega, \qquad 0 \leq s < \infty. \tag{3.13}$$

The following postulate embodies the third of the physical requirements listed above.

Postulate 3. <u>The space $\mathfrak{B}$ contains non-trivial constant functions. That is, for each vector $\Omega$ in $\mathcal{C}$, the function $\Omega^\dagger$ is in</u> C.

It follows from this assumption that given any functional $\mathfrak{f}$ on $\mathfrak{C}$, one can define a function $\mathfrak{f}^\circ$ on $\mathcal{C}$ by the formula

$$\mathfrak{f}^\circ(\Omega) = \mathfrak{f}(\Omega^\dagger) \quad \text{for all } \Omega\in\mathcal{C}; \tag{3.14}$$

$\mathfrak{f}^\circ$ is called the equilibrium response function corresponding to $\mathfrak{f}$.[4] If $\mathfrak{f}$ is a continuous functional on $\mathfrak{C}$, then $\mathfrak{f}^\circ$ is continuous on $\mathcal{C}$.

The norm $\|\cdot\|$ on $\mathfrak{B}$ is said to have the relaxation property,[#]

---

[#]Ref. 18; see also Refs. 16, 17, 4, 13, and 20.

if, for each function $\Phi$ in V,

$$\lim_{\sigma\to\infty} \|\Phi_{(\sigma)} - \Phi(0)^{\dagger}\| = 0, \tag{3.15}$$

where, in accord with (3.13), $\Phi(0)^{\dagger}$ is the constant function on $[0,\infty)$ with value $\Phi(0)$. Clearly, $\|\cdot\|$ has the relaxation property if and only if (3.15) holds for each $\Phi$ in C. Hence the assumption of the relaxation property is equivalent to the assertion that every continuous functional $\mathfrak{g}$ on $\mathfrak{C}$ obeys the relation

$$\lim_{\sigma\to\infty} \mathfrak{g}(\Phi^{(\sigma)}) = \mathfrak{g}(\Phi(0)^{\dagger}) = \mathfrak{g}^{\circ}(\Phi(0)) \tag{3.16}$$

for each $\Phi$ in $\mathfrak{C}$; that is, in the limit of large $\sigma$, the response $\mathfrak{g}(\Phi^{(\sigma)})$ to the static continuation $\Phi^{(\sigma)}$ of an arbitrary history $\Phi$ depends on only the present value of $\Phi$ and is given by the equilibrium response function defined in (3.14).

Postulate 4. The norm $\|\cdot\|$ has the relaxation property.

Postulates 1-4 yield

Theorem 4.# Let $\Lambda_1(\cdot)$ and $\Lambda_2(\cdot)$ be functions mapping $(-\infty,\infty)$ into $\mathcal{V}_{(10)}$ such that, for each t, the histories $\Lambda_1^t$ and $\Lambda_2^t$ are in V. If $\lim_{t\to\infty} |\Lambda_1(t) - \Lambda_2(t)| = 0$, then $\lim_{t\to\infty} \|\Lambda_1^t - \Lambda_2^t\| = 0$.

#Ref. 13, Thm. 5.1.

A function $\Phi$ in C is called a tame history if

($\alpha$) $\Phi$ is differentiable in the classical sense at $s = 0$;

that is,

$$\dot{\Phi}(0) \stackrel{\text{def}}{=} -\frac{d}{ds}\Phi(s)\Big|_{s=0} = \lim_{s\to 0^+}\frac{\Phi(0)-\Phi(s)}{s} \tag{3.17}$$

exists.

($\beta$) $\Phi_r$, the past history of $\Phi$, is an absolutely continuous function on $(0,\infty)$.

($\gamma$) $\mathfrak{B}$ contains an element $\dot{\Phi}$, called the time-derivative of $\Phi$, which obeys the equation

$$\dot{\Phi}(s) = -\frac{d}{ds}\Phi(s), \qquad \mu\text{-a.e.} \tag{3.18}$$

For technical reasons, one assumes

Postulate 5. Tame histories with time-derivatives of compact support are dense in $\mathfrak{C}$. That is, given any $\Psi$ in $\mathfrak{C}$ and any $\delta > 0$, there exists a tame history $\Phi$ in $\mathfrak{C}$ such that $\|\Psi-\Phi\| < \delta$ and $\dot{\Phi}(s) = \underset{\sim}{0}$ for all s outside of a closed bounded set in $[0,\infty)$.

It follows from Postulate 5 that $\mathfrak{B}$ is separable, that continuous functions of compact support are dense in $\mathfrak{B}$, and that $\mathfrak{B}$ has the following dominated-convergence property# familiar in the theory of Lebesgue spaces:

---

#For theorems of this type, see Luxemburg & Zaanen[21], Thm. 2.2, and Luxemburg[22], Thm. 46.2. See also Ref. 13, Remarks 3.1 and 3.2.

*If* $\Psi$ belongs to $\mathfrak{B}$ and if $\Phi_j$ is a sequence of elements of $\mathfrak{B}$ with $|\Phi_j(s)| \leq |\Psi(s)|$, $\mu$-a.e., such that $\Phi_j(s) \to \underset{\sim}{0}$, $\mu$-a.e., *then* $\|\Phi_j\| \to 0$.

Let $\mathfrak{f}$ be a continuous function mapping $\mathfrak{C}$ into a metric space. It follows from Theorem 3 that $\mathfrak{f}$ can be regarded equally well as a function of ordered pairs $(\Phi(0),\Phi_r)$ with $\Phi(0)$ in $\mathcal{C}$ and $\Phi_r$ in $\mathfrak{C}_r$, i.e.

$$\mathfrak{f}(\Phi) = \mathfrak{f}(\Phi(0);\Phi_r), \tag{3.19}$$

and the continuity of $\mathfrak{f}$ over $\mathfrak{C}$ implies that $\mathfrak{f}(\Phi(0);\Phi_r)$ is jointly continuous in the two variables, $\Phi(0)$ in $\mathcal{C}$, and $\Phi_r$ in $\mathfrak{C}_r$. Now, $\mathfrak{f}$ can be used to define a functional transformation mapping functions $\Lambda(\cdot)$ on $(-\infty,\infty)$ into functions $\phi(\cdot)$ on $(-\infty,\infty)$, by setting

$$\phi(t) = \mathfrak{f}(\Lambda^t) \tag{3.20}$$

for each $t\in(-\infty,\infty)$, where $\Lambda^t(s) = \Lambda(t-s)$, $s \geq 0$. Employing Postulates 1, 2, 3, and 5, one can prove that the functional transformation, $\Lambda(\cdot) \mapsto \phi(\cdot)$, preserves regularity in the following sense.

**Theorem 6.**# *Let $\mathfrak{f}$ be a continuous function on $\mathfrak{C}$ with values in a metric space, and suppose that $\Lambda(\cdot)$ is a function on $(-\infty,\infty)$ with $\Lambda^t$ in $\mathfrak{C}$ for each* t. *If $\Lambda(\cdot)$ is a regulated function,* i.e. *a function for which the limits $\lim_{\tau\to t^+}\Lambda(\tau)$ and $\lim_{\tau\to t^-}\Lambda(\tau)$ exist for each* t, *then $\phi(\cdot)$, given by (3.20), is also a regulated function. Furthermore, $\phi(\cdot)$ can suffer discontinuities at only those times $t_i$ at which $\Lambda(\cdot)$ is discontinuous; at all other times $\phi(\cdot)$ is continuous.*

#Ref. 13, Remark 3.3; see also Ref. 18, Remark 5.1.

(To obtain this result one first shows that the mapping $t \rightsquigarrow \Lambda_r^t \in \mathfrak{C}_r$ is continuous, for all t, even for those at which $\Lambda(\cdot)$ experiences a discontinuity.)

Let $\mathfrak{A}$ be a cone in a Banach space $\mathfrak{B}$, and let $\mathfrak{D}$ be the subspace of $\mathfrak{B}$ spanned by $\mathfrak{A}$. A real-valued function $\mathfrak{g}$ defined on $\mathfrak{A}$ is said to be continuously Fréchet-differentiable on $\mathfrak{A}$# if, for each $\underset{\sim}{\phi}$ in $\mathfrak{A}$ and every $\underset{\sim}{\xi}$ in $\mathfrak{B}$ with $\underset{\sim}{\phi}+\underset{\sim}{\xi}$ in $\mathfrak{A}$,

$$\mathfrak{g}(\underset{\sim}{\phi}+\underset{\sim}{\xi}) = \mathfrak{g}(\underset{\sim}{\phi}) + d\mathfrak{g}(\underset{\sim}{\phi}|\underset{\sim}{\xi}) + o(\|\underset{\sim}{\xi}\|), \tag{3.21}$$

where $d\mathfrak{g}(\cdot|\cdot)$ is defined and continuous on $\mathfrak{A}\times\mathfrak{D}$ and is such that $d\mathfrak{g}(\underset{\sim}{\phi}|\underset{\sim}{\xi})$ is a linear function of $\underset{\sim}{\phi}$ for each $\underset{\sim}{\xi}$. The linear functional $d\mathfrak{g}(\underset{\sim}{\phi}|\cdot)$ is called the Fréchet derivative of $\mathfrak{g}$ at $\underset{\sim}{\phi}$.

An argument given in Reference 20 here yields

Theorem 7## (chain rule). If $\mathfrak{g}$ is a real-valued continuously Fréchet-differentiable function on $\mathfrak{C}$, then, for each tame history $\Phi$ in $\mathfrak{C}$, the derivative,

$$\dot{\mathfrak{g}} \overset{\text{def}}{=} \lim_{\sigma\to 0+} \frac{\mathfrak{g}(\Phi)-\mathfrak{g}(\Phi_{(\sigma)})}{\sigma}, \tag{3.22}$$

exists and is given by

$$\dot{\mathfrak{g}} = d\mathfrak{g}(\Phi|\dot{\Phi}), \tag{3.23}$$

with $\dot{\Phi}$ the time-derivative defined in (3.18).

---

#Of course the definition can be employed for other types of subsets of $\mathfrak{B}$, such as open subsets.

##Ref. 20, Remark 1 and Appendix II; see also Refs. 4 and 23. The proof given in Ref. 20 does not require Postulate 4.

Suppose $\mathcal{g}$ is continuously Fréchet-differentiable on $\mathfrak{C}$, and recall that $\mathcal{g}(\Phi)$ can be written

$$\mathcal{g}(\Phi) = \mathcal{g}(\Phi(0);\Phi_r), \tag{3.24}$$

where $\Phi(0)$, in $\mathcal{C}$, is the present value of $\Phi$, and $\Phi_r$, in $\mathfrak{C}_r$, is the restriction of $\Phi$ to $(0,\infty)$. The assumed differentiability of $\mathcal{g}$ on $\mathfrak{C}$ implies the existence, for each $\Phi$, of the <u>instantaneous derivative</u>[4] $D\mathcal{g}(\Phi)$ and the <u>past-history derivative</u> $\delta\mathcal{g}(\Phi|\cdot)$, which are determined by the equations

$$\mathcal{g}(\Phi(0)+\Omega;\Phi_r) = \mathcal{g}(\Phi(0);\Phi_r) + D\mathcal{g}(\Phi)\cdot\Omega + o(|\Omega|) \tag{3.25}$$

and

$$\mathcal{g}(\Phi(0);\Phi_r+\Psi_r) = \mathcal{g}(\Phi(0);\Phi_r) + \delta\mathcal{g}(\Phi|\Psi_r) + o(\|\Psi_r\|_r); \tag{3.26}$$

(3.25) holds for all $\Omega$ in $\mathcal{V}_{(10)}$ with $\Phi(0)+\Omega$ in $\mathcal{C}$, while (3.26) holds for all $\Psi_r$ in $\mathfrak{B}_r$ with $\Phi_r+\Psi_r$ in $\mathfrak{C}_r$. For each $\Phi$ in $\mathfrak{C}$, the value $D\mathcal{g}(\Phi)$ of $D\mathcal{g}$ is a vector in $\mathcal{V}_{(10)}$, and $\delta\mathcal{g}(\Phi|\cdot)$ is a linear function on $\mathfrak{B}_r$. The functionals $D\mathcal{g}$ and $\delta\mathcal{g}$ determine $d\mathcal{g}$ through the relation

$$d\mathcal{g}(\Phi|\Psi) = D\mathcal{g}(\Phi)\cdot\Psi(0) + \delta\mathcal{g}(\Phi|\Psi_r), \tag{3.27}$$

and one can write the chain rule (3.23) in the form

$$\dot{\mathcal{g}} = D\mathcal{g}(\Phi)\cdot\dot{\Phi}(0) + \delta\mathcal{g}(\Phi|\dot{\Phi}_r) \tag{3.28}$$

with $\dot{\Phi}(0)$ the present value, and $\dot{\Phi}_r$ the past history, of $\dot{\Phi}$.

There is now assembled here apparatus sufficient for a precise statement of the principle of fading memory as used in the thermodynamics of simple materials.

<u>Postulate of smoothness for response functionals.</u> <u>For each simple material there exists a history space</u> $\mathfrak{B}$, <u>formed as described above, such that</u>

(1) $\mathfrak{C}$, <u>the cone in</u> $\mathfrak{B}$ <u>corresponding to functions mapping</u> $[0,\infty)$ <u>into</u> $\mathcal{C}$, <u>obeys Postulates</u> 1-5;

(2) <u>the functionals</u> $\mathfrak{p}$, $\mathfrak{S}$, <u>and</u> $\mathfrak{q}$ <u>in</u> (3.6) <u>are defined and continuous on</u> $\mathfrak{C} \times \mathcal{V}_{(3)}$;[#]

(3) <u>the functional</u> $\mathfrak{p}$ <u>is continuously Fréchet-differentiable on</u> $\mathfrak{C} \times \mathcal{V}_{(3)}$.[##]

---

[#] In (3.6), $\Gamma^t \in \mathfrak{C}$, while $\underset{\sim}{g} \in \mathcal{V}_{(3)}$.

[##] $\mathfrak{C} \times \mathcal{V}_{(3)}$ is here considered a cone in $\mathfrak{B} \oplus \mathcal{V}_{(3)}$.

## 4. Thermodynamic Restrictions on Materials with Memory

It is easily shown that under appropriate assumptions of regularity for the dependence of $\underset{\sim}{\chi}$ and $\theta$ upon $\underset{\sim}{\xi}$ and t, it follows from the balance laws (2.1) and (2.2) that the Clausius-Duhem inequality (2.11) can be written in the local form[4]

$$\dot{\psi} \leq \Sigma \cdot \dot{\Gamma} - \frac{1}{\rho\theta} \underset{\sim}{q} \cdot \underset{\sim}{g}. \tag{4.1}$$

Working with this local form of the inequality one can prove the following theorem which gives the restrictions which the second law places on the response functionals $\mathfrak{p}$, $\mathfrak{h}$, $\mathfrak{s}$, and $\mathfrak{q}$ in (2.9) [or, equivalently, $\mathfrak{p}$, $\mathfrak{S}$, and $\mathfrak{q}$ in (3.6)].

Theorem 8.[#] It follows from the Dissipation Principle and the Postulate of Smoothness for Response Functionals that

(i) the functionals $\mathfrak{p}$ and $\mathfrak{S}$ are independent of $\underset{\sim}{g}$; i.e.

$$\psi(t) = \mathfrak{p}(\Gamma^t), \qquad \Sigma(t) = \mathfrak{S}(\Gamma^t), \tag{4.2}$$

whenever $\Gamma^t$ is in $\mathfrak{S}$;

(ii) the functional $\mathfrak{S}$ is determined by the functional $\mathfrak{p}$ through the "generalized stress relation",

$$\mathfrak{S} = D\mathfrak{p}; \tag{4.3}$$

i.e.

$$\Sigma(t) = D\mathfrak{p}(\Gamma^t) \tag{4.4}$$

whenever $\Gamma^t$ is in $\mathfrak{C}$;

---

[#]Ref. 4, Thm. 1, p. 19; see also Ref. 20, Thm. 1.

(iii) <u>for each tame history</u> $\Gamma^t$ <u>in</u> $\mathfrak{C}$

$$\delta \mathfrak{p}(\Gamma^t | \dot{\Gamma}^t) \leq 0 \tag{4.5}$$

<u>and</u>

$$\mathfrak{q}(\Gamma^t; \underset{\sim}{g}) \cdot \underset{\sim}{g} \leq -\rho\theta\delta\mathfrak{p}(\Gamma^t | \dot{\Gamma}^t). \tag{4.6}$$

<u>Furthermore,</u> (i), (ii), <u>and</u> (iii), <u>when taken together, give not only a necessary, but also a sufficient condition on</u> $\mathfrak{p}$, $\mathfrak{s}$, <u>and</u> $\mathfrak{q}$ <u>for</u> (4.1) <u>to hold for all</u> $\underset{\sim}{g}$ <u>in</u> $\mathcal{V}_{(3)}$ <u>and all tame histories</u> $\Gamma^t$ <u>in</u> $\mathfrak{C}$.

When I gave this theorem in my essay[4] of 1964, I proved it using a form of the principle of fading memory less general than that described here. The form used was drawn from earlier work done in collaboration with Walter Noll[14,15,16]. In the present terminology one can say that I employed the Postulate of Smoothness stated at the end of Section 3, but employed for $\mathfrak{B}$ a special type of history space, namely a Hilbert space $\mathfrak{H}$ formed from functions $\Phi$, mapping $[0,\infty)$ in $\mathcal{V}_{(10)}$, for which

$$\|\Phi\|^2 \overset{\text{def}}{=} |\Phi(0)|^2 + \int_0^\infty |\Phi(s)|^2 k(s)ds \tag{4.7}$$

is finite; $k(\cdot)$ was a fixed, positive, monotone-decreasing function, assumed to be summable on $(0,\infty)$, and called the "influence function".[#] Later, Victor Mizel and I[20] observed that Theorem 8 remains valid in the present more general theory.

---

[#] Cf. Refs. 14-16. No further assumptions on k are needed for the main theorems of Ref. 4. See Ref. 18 for an axiomatic approach to history spaces obeying (4.7).

The conclusions (i), (ii), and (iii) of Theorem 8 have some interesting consequences:

Theorem 9.[#] Of all total histories ending with a given value of $\Gamma = (\underset{\sim}{F}, \theta)$, that corresponding to constant values of $\Gamma$ for all times has the least free energy; i.e. for each $\Gamma^t$ in $\mathfrak{C}$

$$\mathfrak{p}^\circ(\Gamma^t(0)) \le \mathfrak{p}(\Gamma^t). \tag{4.8}$$ [##]

Theorem 10.[###] If $\Gamma$ is a vector in $\mathcal{C}$, and if $\Gamma^\dagger$ is the constant function defined by $\Gamma^\dagger(s) \equiv \Gamma$, then for all functions $\Phi_r$ in $\mathfrak{B}_r$,

$$\delta \mathfrak{p}(\Gamma^\dagger | \Phi_r) = 0. \tag{4.9}$$

Theorem 11.[####] The equilibrium response functions corresponding to $\mathfrak{p}$ and $\mathfrak{S}$ obey the classical relation

$$\mathfrak{S}^\circ(\Gamma) = \nabla\mathfrak{p}^\circ(\Gamma) \quad \text{for all } \Gamma \text{ in } \mathcal{C}, \tag{4.10}$$

where $\nabla\mathfrak{p}^\circ$ is the ordinary gradient of $\mathfrak{p}^\circ$.

---

[#] Ref. 4, Thm. 3, p. 26; see also Ref. 20, Thm. 2.

[##] See Eq. (3.14) for the definition of $\mathfrak{p}^\circ$, the equilibrium response function corresponding to $\mathfrak{p}$.

[###] Ref. 4, Corollary to Thm. 3., p. 26; Ref. 18, Thm. 3.

[####] Ref. 4, Remark 11, p. 27; see also Ref. 20, Thm. 4.

Although the proof of Theorem 8 does not employ Postulate 4, the proofs of Theorems 9, 10, and 11 do. For further discussion of this point, see Reference 10 and the papers referred to therein.

## 5. Brief Summary

In his essay, "A Method of Graphical Representation of the Thermodynamic Properties of Substances by Means of Surfaces," published in 1873, Gibbs[24] proposed and studied a criterion for the stability of a fluid surrounded by a medium held at fixed temperature and pressure. Assuming that the specific internal energy $\epsilon$ of a fluid in thermodynamic equilibrium is given by a function $\bar{\epsilon}$ of the specific entropy $\eta$ and specific volume $\upsilon$ of the fluid, Gibbs gave a heuristic argument to the effect that a uniform phase with specific entropy $\eta^\circ$ and specific volume $\upsilon^\circ$ is stable in an environment at temperature $\theta^\circ$ and pressure $p^\circ$ if the inequality

$$\bar{\epsilon}(\eta,\upsilon) - \theta^\circ\eta + p^\circ\upsilon > \bar{\epsilon}(\eta^\circ,\upsilon^\circ) - \theta^\circ\eta^\circ + p^\circ\upsilon^\circ \tag{5.1}$$

holds for all pairs $(\eta,\upsilon)$ not equal to $(\eta^\circ,\upsilon^\circ)$.# Criteria of this type occur also in Gibbs' memoir "On the Equilibrium of Heterogeneous Substances"[25] and are there related to his concept of the stability of an isolated system.

---

#If (5.1) holds with $>$ replaced by $\geq$ and reduces to an equality for some pair $(\eta,\upsilon) \neq (\eta^\circ,\upsilon^\circ)$, then there are two or more stable uniform phases possible at the temperature $\theta^\circ$ and pressure $p^\circ$. A beautiful statical theory of coexistent phases was developed at length by Gibbs. In the present lecture, however, I shall try to avoid the formidable mathematical problems which arise in any attempt to formulate a dynamical theory of coexistent phases.

This lecture, which draws heavily on an essay[29] recently published in the Archive for Rational Mechanics and Analysis, is concerned with the dynamical significance of the criterion (5.1) for stability. I shall here discuss the following question: <u>If a uniform equilibrium state of a fluid body is stable according to the definitions of classical thermostatics, and if the body is, in some precise sense, in a "fixed environment", is it then true that every thermodynamic process in the body which passes near to the equilibrium state at one time must remain near to that state at all future times</u>?

I shall attempt to show that the question has a positive answer for a broad class of materials called "regular fluids".# Each regular fluid has an equilibrium response function $\bar{\epsilon}$ which for states of equilibrium gives $\epsilon$ as a function of $\eta$ and $\upsilon$,

$$\epsilon = \bar{\epsilon}(\eta,\upsilon), \tag{5.2}$$

but which outside of equilibrium yields only a lower bound for $\epsilon$:

$$\epsilon \geq \bar{\epsilon}(\eta,\upsilon). \tag{5.3}$$

For many materials, including simple fluids with fading memory, (5.3) is a consequence of the second law of thermodynamics.## The class of regular

#Cf. Coleman & Greenberg[26] and Coleman[29].

##For simple fluids with fading memory, the relation (5.3) is <u>equivalent</u> to Theorem 9 on page 24. [Cf. Ref. 4, Remark 22, p. 35.]

fluids includes not only fluids with memory, but also the perfect fluids and linearly viscous fluids of classical hydrodynamics.

We shall see that if the equilibrium response function $\bar{\epsilon}$ for a regular fluid body $\mathcal{B}$ is convex outside a compact set, i.e. if the points in the domain $\underline{\underline{D}}$ of $\bar{\epsilon}$ which are not points of convexity for $\bar{\epsilon}$ are contained in a compact subset of $\underline{\underline{D}}$, then whenever $(\eta^\circ, \upsilon^\circ)$ is such that (5.1) holds for all $(\eta, \upsilon) \neq (\eta^\circ, \upsilon^\circ)$, the uniform equilibrium state at $(\eta^\circ, \upsilon^\circ)$ is dynamically stable in the following sense: If a thermodynamic process $\mathcal{T}$ of $\mathcal{B}$ is compatible with immersion of $\mathcal{B}$ in an environment at temperature $\theta^\circ$ and pressure $p^\circ$ and is such that the fields over $\mathcal{B}$ describing the spatial distribution of internal energy, kinetic energy, specific volume, and entropy are close, in $\mathcal{L}_1$, to the corresponding uniform equilibrium fields at some time t, then in the process $\mathcal{T}$ these fields remain close, in $\mathcal{L}_1$, to the equilibrium fields at <u>all</u> times after t. This is the content of Theorem 13, given in Section 9. This theorem and the closely related Theorem 12 follow, by easy arguments, from Remark 2 (see Section 8), which, in turn, rests on a lemma from the theory of functions with points of convexity.

## 6. Thermodynamic Processes in Fluids

Here a <u>body</u> $\mathcal{B}$ is a smooth manifold, of material points X, imbedded in a fixed Euclidean space $\mathcal{E}$ and possessing an intrinsic mass-measure $m$. Of course, a <u>thermodynamic process</u> of $\mathcal{B}$ is a collection $\mathcal{T}$ of functions, of X and the time t, compatible with the laws of balance of momentum and energy. At the present level of generality, each thermodynamic process consists of six functions: (1) the motion $\underset{\sim}{\chi}$, with $\underset{\sim}{x} = \underset{\sim}{\chi}(X,t)$; (2) the temperature $\theta$, (3) the stress tensor $\underset{\sim}{T}$, (4) the specific internal energy $\epsilon$, (5) the specific entropy $\eta$, and (6) the heat flux vector $\underset{\sim}{q}$. It is here assumed that gravitational body forces are absent and that there is no heat transfer by radiation, (such was not the case in Section 2), and hence the laws of balance of momentum and balance of energy take the forms [cf. (2.1) and (2.2)],

$$\frac{d}{dt}\int_{\mathcal{P}} \underset{\sim}{\dot{x}}\, dm = \int_{\partial\mathcal{P}} \underset{\sim}{T}\underset{\sim}{n}\, da, \qquad (6.1)$$

$$\frac{d}{dt}\int_{\mathcal{P}}\left(\epsilon + \frac{1}{2}\dot{x}^2\right)dm = \int_{\partial\mathcal{P}} (\underset{\sim}{\dot{x}}\cdot\underset{\sim}{T}\underset{\sim}{n} - \underset{\sim}{q}\cdot\underset{\sim}{n})\, da, \qquad (6.2)$$

and must hold at all times t and in all parts $\mathcal{P}$ of $\mathcal{B}$.

Many types of processes are imaginable. The theorems of Section 8 concern processes possible in fluid bodies which from some time, say t = 0, onward are either isolated or immersed in an environment held at fixed temperature and pressure.

Definition 1. *If a thermodynamic process $\mathcal{T}$ of a fluid body $\mathcal{B}$ is such that for some positive pair* $(\theta°,p°)$,

$$(\theta-\theta°)\underset{\sim}{q}\cdot\underset{\sim}{n} \geq 0 \quad \text{for all } X\epsilon\partial\mathcal{B} \text{ and all } t \geq 0 \tag{6.3}$$

*and*

$$\int_{\partial\mathcal{B}} \dot{\underset{\sim}{x}}\cdot\underset{\sim}{T}\underset{\sim}{n}\, da = -\dot{V}p° \quad \left(\text{with } V = \int_{\mathcal{B}} \upsilon\, d\mathfrak{m}\right), \text{ for all } t \geq 0, \tag{6.4}$$

*then $\mathcal{T}$ is said to be compatible with immersion of $\mathcal{B}$ in an environment at temperature $\theta°$ and pressure $p°$, from time zero onward.*

The condition (6.3) asserts that if a material point on the surface of $\mathcal{B}$ has a temperature $\left\{\begin{matrix}\text{greater}\\ \text{less}\end{matrix}\right\}$ than $\theta°$, then heat does not flow $\left\{\begin{matrix}\text{into}\\ \text{out of}\end{matrix}\right\}$ $\mathcal{B}$ at that point. This condition is met, trivially, in the extreme cases in which (1) $\mathcal{B}$ is thermally isolated, $(\underset{\sim}{q}\cdot\underset{\sim}{n} \equiv 0)$, and (2) the surface of $\mathcal{B}$ is held at the constant temperature $\theta°$.# If $\mathcal{B}$ is in an environment which is held at the constant and uniform temperature $\theta°$, then (6.3) is expected to be a consequence of the second law of thermodynamics, under reasonable constitutive assumptions for the transfer of heat across the surface of $\mathcal{B}$. For example, if one assumes that for each point on the surface of $\mathcal{B}$,

$$\underset{\sim}{q} = (\theta-\theta°)k\underset{\sim}{n}, \qquad k = k(\theta,\upsilon),$$

then the second law requires that k be not negative, and (6.3) is obeyed.

---

#Cf. Koiter[27].

The condition (6.4) asserts that $Vp^\circ$ is a potential for the total work done by the contact forces acting on $\mathcal{B}$. This condition is met, for example, when $\mathcal{B}$ is mechanically isolated, i.e. when $\dot{V} = 0$ and $\dot{\underset{\sim}{x}}\cdot\underset{\sim}{T}\underset{\sim}{n} = 0$ on the surface of $\mathcal{B}$; a special case of a process compatible with mechanical isolation is one in which $\mathcal{B}$ fills a rigid container to which it adheres, so that the velocity $\dot{\underset{\sim}{x}}$ is zero at the bounding surface of $\mathcal{B}$. Processes which are not compatible with isolation, but yet obey (6.4), occur in a body confined in a piston chamber, provided that the force F exerted by the piston on $\mathcal{B}$ is positive and constant in time; then $p^\circ = F/A$, with A the cross-sectional area of the piston.

The Clausius-Duhem inequality [(2.10) and (2.11)] here reduces to the assertion that, at each instant of time and for each part $\mathcal{P}$ of $\mathcal{B}$,

$$\frac{d}{dt}\int_{\mathcal{P}} \eta \, d\mathcal{m} \;\geq\; -\int_{\partial\mathcal{P}} \frac{1}{\theta}\, \underset{\sim}{q}\cdot\underset{\sim}{n}\, da. \tag{6.5}$$

Given two positive numbers $\theta^\circ$, $p^\circ$, and a process $\mathcal{J}$ of $\mathcal{B}$, one can define, at each time t, a number $\Phi(t)$ by[#]

$$\Phi(t) \;=\; \int_{\mathcal{B}} \left[\epsilon(X,t) - \theta^\circ\eta(X,t) + p^\circ\upsilon(X,t) + \frac{1}{2}\,\dot{x}^2(X,t)\right] d\mathcal{m}. \tag{6.6}$$

I call $\Phi(t)$ the <u>canonical free-energy of</u> $\mathcal{B}$, <u>at time</u> t, <u>under an environment at temperature</u> $\theta^\circ$ <u>and pressure</u> $p^\circ$.

---

[#]Note that $\Phi(t)$ is scalar-valued. The symbol $\Phi$ was used in a completely different sense in Sections 3 & 4.

*Proposition* 1.# It follows from the balance law (6.2) and the Clausius-Duhem inequality (6.5) that if $\mathcal{T}$ is a process of $\mathcal{B}$ compatible with immersion of $\mathcal{B}$ in an environment at temperature $\theta°$ and pressure $p°$ (from time zero onward), and if $\Phi$ is the canonical free-energy of $\mathcal{B}$ under such an environment, then in the process $\mathcal{T}$

$$\frac{d}{dt}\Phi(t) \leq 0,$$

for each $t \geq 0$.

*Proof*. When (6.4) holds, the law of balance of energy, (6.2), reduces to

$$\frac{d}{dt}\int_{\mathcal{B}}\left(\epsilon + \frac{1}{2}\dot{x}^2 + p°\upsilon\right)dm = -\int_{\partial\mathcal{B}} \underset{\sim}{q}\cdot\underset{\sim}{n}\, da. \tag{6.7}$$

Since $\theta$ and $\theta°$ are positive, (6.3) yields

$$-\int_{\partial\mathcal{B}} \underset{\sim}{q}\cdot\underset{\sim}{n}\, da \leq -\theta°\int_{\partial\mathcal{B}}\frac{1}{\theta}\underset{\sim}{q}\cdot\underset{\sim}{n}\, da, \tag{6.8}$$

and the Clausius-Duhem inequality, (6.5), can be written

$$-\theta°\frac{d}{dt}\int_{\mathcal{B}}\eta\, dm \leq \theta°\int_{\partial\mathcal{B}}\frac{1}{\theta}\underset{\sim}{q}\cdot\underset{\sim}{n}\, da. \tag{6.9}$$

Adding (6.7), (6.8), and (6.9), we obtain

$$\frac{d}{dt}\int_{\mathcal{B}}\left(\epsilon + \frac{1}{2}\dot{x}^2 + p°\upsilon - \theta°\eta\right)dm \leq 0. \blacksquare$$

---

#Propositions related to 1 are given by Ericksen[28], Koiter[27], Coleman & Dill[7], and Coleman[29].

For fluids, the concept of a "uniform static state", which plays a central role in classical thermostatics, can be given a meaning in thermodynamics through the following definition.

*Definition* 2. *A uniform static state of a body* $\mathcal{B}$ *is an admissible process of* $\mathcal{B}$ *for which*

$$\eta(X,t) = \eta^\circ = \text{const.},\quad \upsilon(X,t) = \upsilon^\circ = \text{const.},\ \text{and}\ \dot{\underset{\sim}{x}}(X,t) = \underset{\sim}{0}, \tag{6.10}$$

*for all* X *in* $\mathcal{B}$ *and all* t *in* $(-\infty,\infty)$.

The apparatus required to specify with precision the class of materials called "regular fluids" has now been assembled.

*Definition* 3.# *A body* $\mathcal{B}$ *is comprised of a regular fluid if there exists a differentiable function* $\bar{\epsilon}$ *mapping* $(0,\infty)\times(0,\infty)$ *into* $(0,\infty)$ *such that*

(i) *for each pair* $(\eta^\circ,\upsilon^\circ)$ *of positive numbers, there is a unique uniform static state of* $\mathcal{B}$ *with* $\eta = \eta^\circ$ *and* $\upsilon = \upsilon^\circ$;## *in this state (regarded as a process), for all* X *and* t,

---

#This definition is essentially the same as that employed by Coleman & Greenberg[26]. See also Coleman[29].

##*Unique* here means the following: If $\underset{\sim}{\chi}$, $\epsilon$, $\eta$, $\theta$, $\underset{\sim}{q}$, and $\underset{\sim}{T}$ are the six functions of X and t describing an admissible process of $\mathcal{B}$ with $\eta(X,t) \equiv \eta^\circ$, $\upsilon(X,t) \equiv \upsilon^\circ$, and $\partial_t\underset{\sim}{\chi}(X,t) = 0$, then $\epsilon$, $\eta$, $\theta$, $\underset{\sim}{q}$, and $\underset{\sim}{T}$ are completely determined once $\eta^\circ$ and $\upsilon^\circ$ are given, and $\underset{\sim}{\chi}$ is determined to within a constant unimodular transformation, i.e. to within a transformation of the form $\underset{\sim}{\chi}(X,t) \to \underset{\sim}{U}\underset{\sim}{\chi}(X,t) + \underset{\sim}{c}$, with $\underset{\sim}{c}$ a vector and $\underset{\sim}{U}$ a tensor obeying $|\det \underset{\sim}{U}| = 1$.

$$\left.\begin{aligned}
\epsilon(X,t) &= \epsilon^\circ \overset{\text{def}}{=} \bar{\epsilon}(\eta^\circ,\upsilon^\circ),\\
\theta(X,t) &= \theta^\circ \overset{\text{def}}{=} \bar{\theta}(\eta^\circ,\upsilon^\circ), \quad \text{with } \bar{\theta} = \partial_\eta \bar{\epsilon},\\
\underset{\sim}{T}(X,t) &= -p^\circ \underset{\sim}{1}, \quad \text{with } p^\circ \overset{\text{def}}{=} \bar{p}(\eta^\circ,\upsilon^\circ), \quad \bar{p} = -\partial_\upsilon \bar{\epsilon},\\
\underset{\sim}{q}(X,t) &= \underset{\sim}{0};
\end{aligned}\right\} \qquad (6.11)$$

(ii) in <u>every</u> <u>admissible</u> <u>thermodynamic</u> <u>process</u> <u>of</u> $\mathcal{B}$

$$\epsilon(X,t) \geq \bar{\epsilon}(\eta(X,t),\upsilon(X,t)), \qquad (6.12)$$

<u>for all</u> X <u>and</u> t.

<u>The function</u> $\bar{\epsilon}$ <u>in</u> (6.11) <u>and</u> (6.12) <u>is called the</u> *equilibrium energy function* <u>for the regular fluid</u>.

## 7. Preliminary Observations

The ideas introduced in the previous section are brought together in the following remark, which shows that for a regular fluid body $\mathcal{B}$ the inequality (5.1) implies that the canonical free energy under an environment at temperature $\theta^\circ$ and pressure $p^\circ$ has a strict minimum at the uniform static state which gives $\mathcal{B}$ the temperature $\theta^\circ$ and the pressure $p^\circ$.

Remark 1.[29] *Let $\mathcal{B}$ be a regular fluid body, let $(\eta^\circ, \upsilon^\circ)$ be a pair of positive numbers, put*

$$\left.\begin{aligned} \epsilon^\circ &= \bar{\epsilon}(\eta^\circ,\upsilon^\circ), \qquad \theta^\circ = \bar{\theta}(\eta^\circ,\upsilon^\circ), \qquad p^\circ = \bar{p}(\eta^\circ,\upsilon^\circ), \\ \phi^\circ &= \epsilon^\circ - \theta^\circ\eta^\circ + p^\circ\upsilon^\circ, \\ \Phi^\circ &= \int_{\mathcal{B}} \phi^\circ dm = M\phi^\circ, \qquad M = \int_{\mathcal{B}} dm, \end{aligned}\right\} \tag{7.1}$$

*and suppose that the inequality (5.1) holds for every pair $(\eta, \upsilon)$ not equal to $(\eta^\circ, \upsilon^\circ)$. If $\epsilon(X,t)$, $\eta(X,t)$, $\upsilon(X,t)$, and $\dot{\underset{\sim}{x}}(X,t)$ are the values assumed by $\epsilon$, $\eta$, $\upsilon$, and $\dot{\underset{\sim}{x}}$ in an arbitrary admissible process $\mathcal{T}$ of $\mathcal{B}$, and if*

$$\zeta(X,t) = \epsilon(X,t) - \theta^\circ\eta(X,t) + p^\circ\upsilon(X,t), \tag{7.2}$$

*then*

$$\zeta(X,t) \geq \phi^\circ \tag{7.3}$$

*for all X and all t, and consequently, at each time t, for the process $\mathcal{T}$,*

$$\Phi(t) \geq \Phi^\circ, \tag{7.4}$$

*where*

$$\Phi(t) \overset{\text{def}}{=} \int_{\mathcal{B}} \left[\epsilon(X,t) - \theta^\circ \eta(X,t) + p^\circ \upsilon(X,t) + \frac{1}{2}\dot{x}^2(X,t)\right] d$$
$$= \int_{\mathcal{B}} \left[\zeta(X,t) + \frac{1}{2}\dot{x}^2(X,t)\right] dm. \tag{7.5}$$

*Moreover, equality holds in (7.3) only when*

$$\eta(X,t) = \eta^\circ, \quad \upsilon(X,t) = \upsilon^\circ, \quad \text{and} \quad \epsilon(X,t) = \epsilon^\circ,$$

*and hence equality holds in (7.4) only when this condition and the condition $\dot{\underset{\sim}{x}}(X,t) = \underset{\sim}{0}$ hold almost everywhere in $\mathcal{B}$.*

Of course, $\Phi(t)$ is here the canonical free energy of $\mathcal{B}$ in an environment at temperature $\theta^\circ$ and pressure $p^\circ$. The number $\Phi^\circ$ is the canonical free energy of $\mathcal{B}$, under the same environment, in the uniform static state (6.10).

***Proof of Remark* 1.** The only thing that requires proof here is the relation (7.3). Let $\bar{\phi}$ be the function on $(0,\infty) \times (0,\infty)$ defined by

$$\bar{\phi}(\eta,\upsilon) = \bar{\epsilon}(\eta,\upsilon) - \theta^\circ\eta + p^\circ\upsilon \tag{7.6}$$

with $\bar{\epsilon}$ the equilibrium energy function for $\mathcal{B}$. Then

$$\phi^\circ = \bar{\phi}(\eta^\circ,\upsilon^\circ), \tag{7.7}$$

and since, by (6.12),

$$\bar{\epsilon}(\eta(X,t),\upsilon(X,t)) \le \epsilon(X,t),$$

we have

$$\bar{\phi}(\eta(X,t),\upsilon(X,t)) \le \zeta(X,t) \tag{7.8}$$

with equality holding only when $\epsilon(X,t) = \bar{\epsilon}(\eta(X,t),\upsilon(X,t))$. Clearly,

(5.1) implies

$$\bar{\phi}(\eta^\circ,\upsilon^\circ) \leq \bar{\phi}(\eta(X,t),\upsilon(X,t)) \qquad (7.9)$$

with equality holding only when $\eta(X,t) = \eta^\circ$ and $\upsilon(X,t) = \upsilon^\circ$. In view of (7.7) and (7.9), the relation (7.8) yields

$$\phi^\circ \leq \zeta(X,t)$$

with equality holding only when $\eta(X,t) = \eta^\circ$, $\upsilon(X,t) = \upsilon^\circ$, and $\epsilon(X,t) = \bar{\epsilon}(\eta^\circ,\upsilon^\circ) = \epsilon^\circ$. ■

The conclusion (7.4) above is strengthened in Remark 2 below. To formulate and prove that remark, however, I must first state some definitions and results from the general theory of functions with points of convexity.

Let $\underline{\underline{R}}$ be the set of real numbers, let $\underline{\underline{R}}^n$, $n \geq 1$, be the space of n-tuples of real numbers, let $\underline{\underline{D}}$ be a convex open subset of $\underline{\underline{R}}^n$, and let f be a function mapping $\underline{\underline{D}}$ into $\underline{\underline{R}}$. A point $\underset{\sim}{z}$ in $\underline{\underline{D}}$ is called a <u>point of convexity for</u> f if

$$f(\underset{\sim}{z}) \leq \alpha f(\underset{\sim}{x}) + (1-\alpha)f(\underset{\sim}{y}),$$

whenever $\underset{\sim}{x}$, $\underset{\sim}{y}$ in $\underline{\underline{D}}$ and $\alpha$ in $[0,1]$ are such that

$$\alpha\underset{\sim}{x} + (1-\alpha)\underset{\sim}{y} = \underset{\sim}{z}.$$

It is easy to show that, when f is differentiable, $\underset{\sim}{z}$ is a point of convexity for f if and only if, for each $\underset{\sim}{x}$ in $\underline{\underline{D}}$,

$$f(\underset{\sim}{x}) \geq f(\underset{\sim}{z}) + (\underset{\sim}{x}-\underset{\sim}{z})\cdot\nabla f(\underset{\sim}{z})$$

with $\nabla f(\underset{\sim}{z})$ the gradient of f at $\underset{\sim}{z}$. If $\underline{\underline{D}}$ has a compact subset $\underline{\underline{S}}$ such that every point in $\underline{\underline{D}} - \underline{\underline{S}}$ is a point of convexity for f, then f is said to be <u>convex outside a compact set</u>. When $\underset{\sim}{0} = (0,\dots,0)$ is in $\underline{\underline{D}}$, f is said to be <u>positive definite</u> on $\underline{\underline{D}}$ if $f(\underset{\sim}{x}) > 0$ for all $\underset{\sim}{x}$ in $\underline{\underline{D}}$ with $\underset{\sim}{x} \neq \underset{\sim}{0}$.

In the lemma which I now state, a norm $\|\cdot\|$ on $\underline{\underline{R}}^n$ is assumed given. Although the lemma is valid regardless of the choice of this norm, in applications $\|\cdot\|$ is defined by

$$\|\underset{\sim}{y}\| = |y_1| + |y_2| + \cdots + |y_n|,$$

where $\underset{\sim}{y} = (y_1,\dots,y_n)$ and $|y_i|$ is the absolute value of $y_i$.

**Lemma.** <u>Suppose $\underline{\underline{D}}$ is a convex open subset of $\underline{\underline{R}}^n$ containing the point $\underset{\sim}{0}$. Let f be a continuous function mapping $\underline{\underline{D}}$ into $\underline{\underline{R}}$ with $f(\underset{\sim}{0}) = 0$, and let $\mathcal{B}$ be a set with a finite, positive measure $m$. If f is convex outside a compact set and positive definite on $\underline{\underline{D}}$, then for each $\epsilon > 0$ there exists a $\delta > 0$ such that every $m$-measurable function g mapping $\mathcal{B}$ into $\underline{\underline{D}}$ with</u>

$$\int_{\mathcal{B}} f(g(X))\,dm < \delta,$$

<u>obeys</u>

$$\int_{\mathcal{B}} \|g(X)\|\,dm < \epsilon.$$

For a proof of this lemma see Reference 29.

*Remark* 2.[#] For each regular fluid body $\mathcal{B}$ whose equilibrium energy function $\bar{\epsilon}$ is convex outside a compact set, there exists a positive function $\delta(\omega,\eta^\circ,\upsilon^\circ)$ on $(0,\infty)\times(0,\infty)\times(0,\infty)$ such that *if* $\epsilon^\circ$, $\theta^\circ$, $p^\circ$, and $\Phi^\circ$ are given by (7.1), *if* $\eta^\circ$ and $\upsilon^\circ$ are such that (5.1) holds for every pair $(\eta,\upsilon)$ not equal to $(\eta^\circ,\upsilon^\circ)$, and *if* $\mathcal{T}$ is an admissible process of $\mathcal{B}$ with

$$\Phi(t) - \Phi^\circ < \delta(\omega,\eta^\circ,\upsilon^\circ), \tag{7.10}$$

where $\Phi(t)$ is defined by (7.5), *then* in the process $\mathcal{T}$, at time t,

$$\left.\begin{aligned} &\theta^\circ\int_{\mathcal{B}} |\eta(X,t)-\eta^\circ|\,dm < \omega, && p^\circ\int_{\mathcal{B}} |\upsilon(X,t)-\upsilon^\circ|\,dm < \omega, \\ &\frac{1}{2}\int_{\mathcal{B}} \dot{x}^2(X,t)\,dm < \omega, \quad \text{and} && \int_{\mathcal{B}} |\epsilon(X,t)-\epsilon^\circ|\,dm < \omega. \end{aligned}\right\} \tag{7.11}$$

*Proof*. Let $\omega > 0$, $\eta^\circ > 0$, and $\upsilon^\circ > 0$ be given, and let $\epsilon^\circ$, $\theta^\circ$, $p^\circ$, $\Phi^\circ$, $\Phi(t)$, $\phi^\circ$, and $\zeta(X,t)$ be as in Remark 1. Then

$$\Phi(t) - \Phi^\circ = \int_{\mathcal{B}} [\zeta(X,t)-\phi^\circ]\,dm + \frac{1}{2}\int_{\mathcal{B}} \dot{x}^2(X,t)\,dm. \tag{7.12}$$

Suppose now that $\eta^\circ$ and $\upsilon^\circ$ are such that (5.1) holds for all $(\eta,\upsilon) \neq (\eta^\circ,\upsilon^\circ)$. By the conclusion (7.3) of Remark 1,

$$\zeta(X,t) - \phi^\circ \geq 0,$$

and therefore (7.12) yields

$$\Phi(t) - \Phi^\circ \geq \frac{1}{2}\int_{\mathcal{B}} \dot{x}^2(X,t)\,dm. \tag{7.13}$$

Furthermore, since $\dot{x}^2$ is never negative, (7.12) yields also

$$\Phi(t) - \Phi^\circ \geq \int_{\mathcal{B}} [\zeta(X,t)-\phi^\circ]\,dm,$$

[#] Ref. 29, Remark 3.3.

and, by (7.8), this implies

$$\Phi(t) - \Phi^\circ \geq \int_{\mathcal{B}} [\bar{\phi}(\eta,\upsilon) - \phi^\circ]\, dm, \qquad (7.14)$$

where I have written $\eta$ for $\eta(X,t)$ and $\upsilon$ for $\upsilon(X,t)$, and $\bar{\phi}$ is as in (7.6). Let

$$y_1 = (\eta-\eta^\circ)\theta^\circ, \qquad y_2 = (\upsilon-\upsilon^\circ)p^\circ, \qquad \underset{\sim}{y} = (y_1,y_2), \qquad (7.15)$$

$$f(\underset{\sim}{y}) \overset{\text{def}}{=} \bar{\phi}\left(\frac{y_1}{\theta^\circ}+\eta^\circ, \frac{y_2}{p^\circ}+\upsilon^\circ\right) - \phi^\circ = \bar{\phi}(\eta,\upsilon) - \phi^\circ. \qquad (7.16)$$

In terms of the function f, (7.14) becomes

$$\Phi(t) - \Phi^\circ \geq \int_{\mathcal{B}} f(\underset{\sim}{y})\, dm. \qquad (7.17)$$

By (7.16), (7.6), and (7.1),

$$f(\underset{\sim}{y}) = \bar{\epsilon}\left(\frac{y_1}{\theta^\circ}+\eta^\circ, \frac{y_2}{p^\circ}+\upsilon^\circ\right) - \epsilon^\circ + y_1 + y_2, \qquad (7.18)$$

and, because $\bar{\epsilon}$ is defined and differentiable on $(0,\infty)\times(0,\infty)$, f is defined and differentiable for all $\underset{\sim}{y}$ in the set

$$\underset{=}{D} = (-\eta^\circ\theta^\circ,\infty) \times (-\upsilon^\circ p^\circ,\infty).$$

Of course, since $\bar{\phi}(\eta^\circ,\upsilon^\circ) = \phi^\circ$, (7.16) yields $f(\underset{\sim}{0}) = 0$. It follows from (5.1) that f is positive definite on $\underset{=}{D}$. [See (7.9).] As it is here assumed that $\bar{\epsilon}$ is convex outside a compact set, it is clear from (7.18) that f is too. Thus, f meets the hypothesis of the lemma stated above, and there exists a $\lambda > 0$ such that

$$\int_{\mathcal{B}} f(\underset{\sim}{y})\, dm < \lambda \implies \int_{\mathcal{B}} \|y\|\, dm < \frac{\omega}{2}.$$

Hence, if we use the norm $\|\cdot\|$ defined before the statement of the lemma, and note (7.17) and (7.15), we have

$$\Phi(t)-\Phi^\circ < \lambda \implies \theta^\circ\int_{\mathcal{B}} |\eta-\eta^\circ|\,dm + p^\circ\int_{\mathcal{B}} |\upsilon-\upsilon^\circ|\,dm < \frac{\omega}{2}\,. \tag{7.19}$$

It follows from (7.1), (7.5), and (7.3) that

$$\Phi(t) - \Phi^\circ = \int_{\mathcal{B}} [\phi-\phi^\circ]\,dm = \int_{\mathcal{B}} |\phi-\phi^\circ|\,dm, \tag{7.20}$$

where

$$\phi = \epsilon - \theta^\circ\eta + p^\circ\upsilon + \frac{1}{2}\dot{x}^2. \tag{7.21}$$

Since

$$\epsilon - \epsilon^\circ = \phi - \phi^\circ + (\eta-\eta^\circ)\theta^\circ - (\upsilon-\upsilon^\circ)p^\circ - \frac{1}{2}\dot{x}^2,$$

the triangle inequality, (7.20), and (7.13) yield

$$\begin{aligned}\int_{\mathcal{B}} |\epsilon-\epsilon^\circ|\,dm &\le \int_{\mathcal{B}} |\phi-\phi^\circ|\,dm + \theta^\circ\int_{\mathcal{B}} |\eta-\eta^\circ|\,dm + p^\circ\int_{\mathcal{B}} |\upsilon-\upsilon^\circ|\,dm + \frac{1}{2}\int_{\mathcal{B}} \dot{x}^2\,dm \\ &\le 2[\Phi(t)-\Phi^\circ] + \theta^\circ\int_{\mathcal{B}} |\eta-\eta^\circ|\,dm + p^\circ\int_{\mathcal{B}} |\upsilon-\upsilon^\circ|\,dm.\end{aligned} \tag{7.22}$$

It is a consequence of (7.13), (7.19), and (7.22) that if one puts

$$\delta = \min(\lambda, \omega/4),$$

then $\delta$ is positive, and all four of the inequalities in (7.11) hold whenever $\Phi(t) - \Phi^\circ$ is less than $\delta$. Of course the $\delta$ so obtained depends on not only $\omega$ but also $\eta^\circ$ and $\upsilon^\circ$, i.e. $\delta = \delta(\omega,\eta^\circ,\upsilon^\circ)$. The argument just given rests on the assumption that (5.1) holds; if $\eta^\circ$ and $\upsilon^\circ$ are not such that (5.1) holds for all $(\eta,\upsilon) \neq (\eta^\circ,\upsilon^\circ)$, then let $\delta(\omega,\eta^\circ,\upsilon^\circ)$ equal $\omega$. Thus one obtains a function $\delta(\omega,\eta^\circ,\upsilon^\circ)$ with the desired properties. ■

## 8. Dynamical Stability

In the two theorems of this section, a regular fluid body $\mathcal{B}$ is supposed assigned in advance, and, for each pair $(\theta^\circ, p^\circ)$ of positive numbers, $\mathfrak{T}(\theta^\circ, p^\circ)$ denotes the class of admissible thermodynamic processes of $\mathcal{B}$ that are compatible with immersion of $\mathcal{B}$ in an environment at temperature $\theta^\circ$ and pressure $p^\circ$ from time $t = 0$ onward.

Theorem 12.# Let $\theta^\circ$, $p^\circ$, $\eta^\circ$, $\upsilon^\circ$, and $\epsilon^\circ$ be the values of the temperature, pressure, specific entropy, specific volume, and specific internal energy in a uniform static state of a regular fluid body $\mathcal{B}$ for which the equilibrium energy function is convex outside a compact set. If (5.1) holds for every pair $(\eta, \upsilon)$ not equal to $(\eta^\circ, \upsilon^\circ)$, then given any $\omega > 0$, there exists a $\delta = \delta(\omega, \eta^\circ, \upsilon^\circ) > 0$ such that each process in $\mathfrak{T}(\theta^\circ, p^\circ)$ which, at any one time $t \geq 0$, satisfies

$$\Phi(t) - \Phi^\circ < \delta, \tag{8.1}$$

with $\Phi(t)$ and $\Phi^\circ$ defined as in Remark 1, must also satisfy

$$\left.\begin{array}{ll} \frac{1}{2}\int_{\mathcal{B}} \dot{x}^2(X,\tau)\,dm < \omega, & \int_{\mathcal{B}} |\epsilon(X,\tau) - \epsilon^\circ|\,dm < \omega, \\ \theta^\circ \int_{\mathcal{B}} |\eta(X,\tau) - \eta^\circ|\,dm < \omega, & \text{and} \quad p^\circ \int_{\mathcal{B}} |\upsilon(X,\tau) - \upsilon^\circ|\,dm < \omega, \end{array}\right\} \tag{8.2}$$

for all $\tau \geq t$.

---

#Ref. 29, Thm. 3.1.

*Proof*. Let $\delta = \delta(\omega,\eta^\circ,\upsilon^\circ)$ be as in Remark 2. Of course, $\delta$ is then always positive. Now, according to Proposition 1, for $\tau \geq 0$, $\Phi(\tau)$ does not increase with $\tau$ in any process $\mathcal{T}$ belonging to $\mathfrak{T}(\theta^\circ,p^\circ)$. Hence if $\mathcal{T}$ in $\mathfrak{T}(\theta^\circ,p^\circ)$ obeys (8.1) at some time $t \geq 0$, then for the process $\mathcal{T}$,

$$\Phi(\tau) - \Phi^\circ < \delta, \tag{8.3}$$

at each time $\tau \geq t$. But, by Remark 2, if (8.3) holds at time $\tau$, then (8.2) also holds at time $\tau$. ■

*Theorem* 13.[#] Let $\theta^\circ$, $p^\circ$, $\eta^\circ$, $\upsilon^\circ$, and $\epsilon^\circ$ be the values of the temperature, pressure, specific entropy, specific volume, and specific internal energy in a uniform static state of a regular fluid body $\mathcal{B}$ for which the equilibrium energy function $\epsilon$ is convex outside a compact set. Suppose (5.1) holds for every pair $(\eta,\upsilon)$ not equal to $(\eta^\circ,\upsilon^\circ)$. Then, given any $\omega > 0$, there exists a $\lambda = \lambda(\omega,\theta^\circ,p^\circ) > 0$ such that if a process $\mathcal{T}$ in $\mathfrak{T}(\theta^\circ,p^\circ)$ has

$$\left.\begin{array}{ll} \frac{1}{2}\int_{\mathcal{B}} \dot{x}^2(X,t)\,dm < \lambda, & \int_{\mathcal{B}} |\epsilon(X,t)-\epsilon^\circ|\,dm < \lambda, \\ \theta^\circ\int_{\mathcal{B}} |\eta(X,t)-\eta^\circ|\,dm < \lambda, \quad \text{and} & p^\circ\int_{\mathcal{B}} |\upsilon(X,t)-\upsilon^\circ|\,dm < \lambda, \end{array}\right\} \tag{8.4}$$

at any one time $t \geq 0$, then $\mathcal{T}$ must obey

$$\left.\begin{array}{ll} \frac{1}{2}\int_{\mathcal{B}} \dot{x}^2(X,\tau)\,dm < \omega, & \int_{\mathcal{B}} |\epsilon(X,\tau)-\epsilon^\circ|\,dm < \omega, \\ \theta^\circ\int_{\mathcal{B}} |\eta(X,\tau)-\eta^\circ|\,dm < \omega, \quad \text{and} & p^\circ\int_{\mathcal{B}} |\upsilon(X,\tau)-\upsilon^\circ|\,dm < \omega, \end{array}\right\} \tag{8.5}$$

for *all* $\tau \geq t$.

[#] Ref. 29, Thm. 3.2.

Proof. It is clear from Remark 1 [see also (7.20)], that in the process $\mathcal{T}$,

$$\Phi(t) - \Phi^\circ = \int_{\mathcal{B}} |\phi(X,t) - \phi^\circ| \, d\mathfrak{m}$$

where

$$\phi(X,t) - \phi^\circ = \frac{1}{2}\dot{x}^2(X,t) + \epsilon(X,t) - \epsilon^\circ - [\eta(X,t) - \eta^\circ]\theta^\circ + [\upsilon(X,t) - \upsilon^\circ]p^\circ.$$

Hence, the triangle inequality yields

$$\Phi(t) - \Phi^\circ \leq \int_{\mathcal{B}} \frac{1}{2}\dot{x}^2(X,t)\, d\mathfrak{m} + \int_{\mathcal{B}} |\epsilon(X,t) - \epsilon^\circ| \, d\mathfrak{m} + \theta^\circ \int |\eta(X,t) - \eta^\circ| \, d\mathfrak{m} + p^\circ \int |\upsilon(X,t) - \upsilon^\circ| \, d\mathfrak{m}$$

and, if we put

$$\lambda = \lambda(\omega, \theta^\circ, p^\circ) = \frac{1}{4}\delta(\omega, \eta^\circ, \upsilon^\circ)$$

with $\delta(\omega, \eta^\circ, \upsilon^\circ)$ as in Theorem 12, then $\lambda$ is not only positive but is also such that (8.4) implies (8.1). But, Theorem 12 asserts that if $\mathcal{T}$ satisfies (8.1) at some time $t \geq 0$, then $\mathcal{T}$ must obey (8.5) for all $\tau \geq t$. ∎

## References


1. B. D. Coleman and W. Noll, Arch. Rational Mech. Anal. **13**, 167 (1963).
2. B. D. Coleman and V. J. Mizel, J. Chem. Phys. **40**, 1116 (1964).
3. B. D. Coleman and M. E. Gurtin, J. Chem. Phys. **47**, 597 (1967).
4. B. D. Coleman, Arch. Rational Mech. Anal. **17**, 1, 230 (1964).
5. B. D. Coleman and M. E. Gurtin, Arch. Rational Mech. Anal. **19**, 266, 317 (1965).
6. B. D. Coleman and V. J. Mizel, Arch. Rational Mech. Anal. **29**, 105 (1968); **30**, 173 (1968).
7. B. D. Coleman and E. H. Dill, Arch. Rational Mech. Anal. **30**, 197 (1968).
8. D. R. Owen, Arch. Rational Mech. Anal. **31**, 91 (1968).
9. D. R. Owen and W. O. Williams, Arch. Rational Mech. Anal. **33**, 288 (1969).
10. B. D. Coleman and D. R. Owen, Arch. Rational Mech. Anal. **36**, 245 (1970).
11. W. Noll, Arch. Rational Mech. Anal. **2**, 197 (1958); **27**, 1 (1967).
12. C. Truesdell and R. A. Toupin, The Classical Field Theories, in the Encyclopedia of Physics, Vol. III/1, Springer-Verlag, Heidelberg (1960).
13. B. D. Coleman and V. J. Mizel, Arch. Rational Mech. Anal. **29**, 18 (1968).
14. B. D. Coleman and W. Noll, Arch. Rational Mech. Anal. **6**, 355 (1960).
15. B. D. Coleman and W. Noll, Rev. Mod. Phys. **33**, 239 (1961); errata **36**, 1103 (1964).
16. B. D. Coleman and W. Noll, Proc. Intl. Sympos. Second-Order Effects, Haifa, 530 (1962).

17. C.-C. Wang, Arch. Rational Mech. Anal. **18**, 117 (1965).

18. B. D. Coleman and V. J. Mizel, Arch. Rational Mech. Anal. **23**, 87 (1966).

19. C. Truesdell and W. Noll, The Non-Linear Field Theories of Mechanics, in the Encyclopedia of Physics, Vol. III/3, Springer-Verlag, Heidelberg (1965).

20. B. D. Coleman and V. J. Mizel, Arch. Rational Mech. Anal. **27**, 255 (1967).

21. W. A. J. Luxemburg and A. C. Zaanen, Math. Annalen **149**, 150 (1963); **162**, 337 (1966).

22. W. A. J. Luxemburg, Indag. Math. **27**, 229 (1965).

23. V. J. Mizel and C.-C. Wang, Arch. Rational Mech. Anal. **23**, 124 (1966).

24. J. W. Gibbs, Trans. Conn. Acad. **2**, 382-404 (1873) = The Scientific Papers of J. Willard Gibbs, Vol. I, pp. 33-54, Longmans-Green, London, New York, 1906.

25. J. W. Gibbs, Trans. Conn. Acad. **3**, 1-8-248 (1875-8) = The Scientific Papers, Vol. I, pp. 55-353.

26. B. D. Coleman and J. M. Greenberg, Arch. Rational Mech. Anal. **25**, 321 (1967).

27. W. T. Koiter, Report 360, Laboratory of Engineering Mechanics, Technological University, Delft (1967).

28. J. L. Ericksen, Int. J. Solids Structures **2**, 573 (1966).

29. B. D. Coleman, Arch. Rational Mech. Anal. **36**, 1 (1970).

CONTENTS

MIX
Papier aus verantwortungsvollen Quellen
Paper from responsible sources
FSC® C105338

If you have any concerns about our products,
you can contact us on
**ProductSafety@springernature.com**

In case Publisher is established outside the EU,
the EU authorized representative is:
**Springer Nature Customer Service Center GmbH**
**Europaplatz 3, 69115 Heidelberg, Germany**

Printed by Libri Plureos GmbH
in Hamburg, Germany